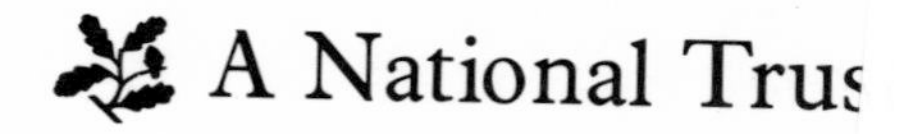

Boat Boy

THE INSIDE STORY OF CANALS

Harry T Sutton

BATSFORD – HERITAGE BOOKS

Illustrations: Chapter 1, Bob Cox; Chapter 2, Vivian Turner; Chapter 3, Moira Clinch

Produced by Heritage Books, 36 Great Russell Street, London WC1B 3PP

Published jointly by B T Batsford Limited and Heritage Books

Distributed by B T Batsford Limited, 4 Fitzhardinge Street, London W1H 0AH

Made and printed in Great Britain by William Clowes & Sons Limited, Beccles, Suffolk

ISBN 0 7134 2121 5

Contents

1 The Jenny

It was in 1830, or thenabout, that I ran away.

'You're ten years old, turning eleven, Will,' my father told me.

''Tis time you earned a living for yourself, lad.' And I was given to my old grandpa to live in his boat.

We were a canal family, the Rowlands. A cut above the ordinary run for we owned our own boats. Grandad's was called *Swallow* and father's was the *Polly*, which was my mother's name. We all lived on board and that was why I had to leave home. There were three of us children already and my mother was soon to have another baby which would have meant six of us, crowded together into a cabin not big enough to swing a cat. But even though I knew I must go, the thought of leaving home brought tears to my eyes.

'Can the lad work locks?' asked my Grandpa, his horny old hand squeezing my thin arms to see if I had muscle enough.

'Not yet, Grandad,' said my father. 'But he be a great one with the horse.'

'Huh,' snorted the old man. 'You've been too soft with him, that's fer sure. Why when I was a littler lad than him I could lift a hunderdweight sack of coals—and steer the boat with t'other hand!'

'The boy'll strengthen if you feed him right, Grandad,' said my mother and I could see that my going made her very sad.

But that was that. I was grandfather's lad from then on and there was no way out. Leastways, at that moment there seemed not to be. Both boats were in the canal basin at Stratford-upon-Avon when I went off in the *Swallow* and I felt very lonely as Grandpa steered the boat away. Grandpa and my father picked up cargoes anywhere along the canals where they could be found and the two boats might not meet again for a year. But unexpected things do happen and I was not to stay long in the *Swallow* as you will hear.

My Grandma had died a few years past, leaving Grandpa alone and to help him with the boat he had Silas Jones, a big brawny lad, about twice my age and three times as strong. Silas took a dislike to me at once so far as I could see, for he took every opportunity to make life miserable from the moment I came aboard.

'Come here, Shiner!' he would shout. He nicknamed me 'Shiner' on account of my duty to shine all the brass fittings in the boat, a job I hated from the start. 'Here, Shiner! Take this rope and look sharp.' And if I fumbled it or dropped the rope in the canal, he would take a swipe with the wet end across my back, enough sometimes to break my tender skin and draw blood.

He was a bully right enough and took every chance to torment me. But Grandpa was too old to bother himself about my bad treatment; or perhaps he was a bit afraid of Silas Jones himself.

In any event, the only friend I had during my time with the *Swallow* was Jenny, the barge horse, and I spent as much time on the tow path with her as I possibly could.

My father was right when he said I was good with a horse. It came naturally I suppose, but no sooner did I see a horse and that horse see me than we were friends. That is how it was with Grandpa's horse. We were friends at once.

Poor Jenny was thin as a rake for Silas rarely fed her and Grandad was too old to bother. But I saw at once she was no ordinary mare. Her head was light and lean, more like that of a racehorse or hunter than a horse for pulling loads. And there was something about the way she stood; not slouched and tired-looking like most canal horses but upright and proud as though she knew she was really intended for better things. The poor animal grazed as well as she could along the grass verges of the canal but there was never enough fodder for a horse with a 30

ton boat to pull.

I used to creep out at night when we were tied up beside the canal bank and climb through hedges into nearby farmers' fields to find food for Jenny. Often she was lucky when I ran across some sweet carrots or a patch of turnips (she loved those!) but more often I could only cut grass from the ditches and what she really needed, of course, was a good feed of maize or oats.

But now we come to the day I ran away. The *Swallow* was loaded with a cargo of grain from the farms round Stratford which we were taking to the cornmarket in Birmingham. There are more than fifty locks on the Stratford Canal which made our journey very slow. If Grandpa could manage it we always stopped for the night near an alehouse and there were dozens of them in those days along all the canals; most of them just cottages with brewing houses behind where the canal boatmen could drink. Grandpa and Silas used to go off together, leaving me to look after the boat and, of course, they always arrived back drunk.

I didn't mind this so long as they just went to bed and left me alone, but this was not good enough for Silas. One night, we were stopped near a place along the canal called Hockley Heath and Silas had gone with Grandad to the Wharf Inn to drink. When they arrived back, very late that night, Silas woke me up.

'Get up out of there, Shiner,' he shouted, his voice thick with drink.

'Help me off with these b.... boots!'

He had the heavy, ironshod boots worn by all boatmen in those days. They had sharp metal studs in the soles to get a grip on the walls when they 'legged' boats through tunnels. I tried to help Silas with his boots, for he was too drunk to help himself, but they were too heavy for me.

'Get away then you miserable creature,' he yelled at me. 'I'll get the b.... things of meself!' and he kicked me away with a vicious jab of his heavy boot. The blow hit me in the face so that I was for a moment knocked unconscious. Silas laughed when he saw the blood streaming down my face.

'That'll teach you, young Shiner!' he told me. Then climbing

into his bunk in the tiny cabin, he fell into a drunken sleep. Grandpa, by this time, was also fast asleep so I alone was awake, my face hurting badly and with the whole night to spend in pain.

It was then I decided to run away, for I could see that one day Silas would certainly kill me in one of his drunken bouts. Finding a short rope to act as a rein, I took Jenny's bit and bridle from its hook and crept quietly ashore to where the mare was tethered for the night. I quickly fitted her harness and in a moment was on her broad, safe back. Then, taking a last look at *Swallow* lying in silence beside the bank, I rode off along the towpath into the night.

My first thought was to retrace the way we had come; back

towards Stratford to look for my family. But then, I realised, father would have to send me back to the *Swallow*—with probably a beating into the bargain for running away. So I decided instead to ride north along the towpath and look for work upon another boat. One owned by somebody less cruel than Silas and Grandpa. I rode for several hours until at last I was so tired and Jenny was going so slowly that I simply had to stop or fall off in my sleep.

Fortunately there was a full moon and turning off the towpath down a lane I was able to find a gate leading into a field of young clover. Opening the gate I turned Jenny loose to graze and then finding a strawstack near the hedge, I settled myself down and went to sleep.

It was broad daylight when I awoke. When I sat up I heard a startled cry and leaping to my feet I prepared to run, but when I saw the cause of my alarm my fear turned to relief. A girl of about my own age was staring fearfully at me from beside the strawrick, a canvas bag clutched in her arms.

'I'll give it back to you, honest I will!' she said, and when I

moved towards her, she began to run away.

'Stop!' I shouted. 'Don't go, I want to speak to you!' When she stopped and looked round, I was quickly by her side and could see that her canvas bag was full of fresh picked clover. I grinned when I saw it.

'For your boat horse?' I asked her. She nodded shyly.

'Don't worry,' I told her. 'I'm not a farmer's boy. I'm a boat boy and that's my horse.' I pointed to Jenny, contentedly grazing in the field.

'You'd best get her out of here then,' she told me. 'My dad says the farmer is a right tartar that owns this field. He told me to keep a sharp lookout.' She looked curiously at me. 'That is why I was skeered—and your face too, that frightened me.' And then I remembered. There must still be blood on my face from that savage kick.

'Can I come back with you to your boat?' I asked her. She nodded.

'My mum will see to your hurt face,' she said. And catching Jenny, we went off together to the canal.

Mary Fisher, for that was her name, was a good friend to me. She took me to *Blackbird,* her family boat which was tied up at the canal side at the end of the lane and without asking any questions about how I was hurt, Mrs. Fisher had soon washed the blood off my face and put some ointment on the broken skin. Then I was given breakfast—cold boiled beef and bread—which tasted like roast goose to me for I was ravenous after my long night's ride. It was not until we had eaten that Mary's father asked any questions. Then he said, very quietly, pointing to Jenny:

'Isn't that Grandpa Rowlands' horse?'

'Yes, sir,' I told him. And I had to think fast to explain how she came to be there. 'I'm Will Rowlands, and Grandpa has just died.' Boatmen are suspicious-minded people and it was not easy to make Mr. Fisher believe that Jenny was really mine. But when you begin to tell lies, one lie seems to lead to another until you get caught up in a tangle of them. And that is what now happened to me. I was soon describing how the *Swallow* was rammed by another boat in a lock and how it sank with

both Silas and Grandad on board so that they were both drowned. And how I dived in to rescue them and hurt my face against the lock gate. Then how, the only one left alive, I had to go off with Grandpa's horse, to look for a job.

I was never sure if my story was believed or whether, knowing Grandpa and Silas and their ways, they guessed the truth. But believe me or not, the Fishers took me into their boat and harnessing Jenny behind their own horse, we set off up the canal to Birmingham.

The next few days were probably the happiest of my life. When a danger is passed, the feeling of safety that follows is twice as sweet. Gliding along on board *Blackbird* through the long summer days, life was all pleasure for me. There was work to be done, of course, helping Mary with the heavy lock gates; taking a quant to shaft the boat into position; jumping ashore to loop a rope round a bollard to steady her in the lock as the water rose, for it is all uphill to Birmingham along the Stratford Canal. Mary was every bit as strong as me and together we were able to work the locks, turning windlasses and swinging gates, whereas one of us alone could not have managed at all.

I think I must have fallen a bit in love with Mary. At any rate, when the time came to say goodbye, I felt really miserable and I daresay Mary was sorry, too. It came about like this.

The *Blackbird* was a 'monkey boat', a narrow boat that is, and a slow one at that. She carried coal, bricks, and farm goods which were never wanted in a hurry so that Mary and her family could take their time along the cut, stopping every night and swimming along at a steady pace all day. They never moved on Sundays or special days like Christmas, Easter or Whit. But not all canal boats were so slow.

Blackbird had just gone through King's Norton tunnel, Mary's father and a legger walking it through whilst Mary and I took the horses over the top, when we were overtaken by a fly-boat, an express one that is, which had right of way. These fast boats were worked by pairs of horses, changed for fresh ones at stables every 3 or 4 miles along the canalside. Slow boats like ours had to make way for them and as this boat and its team came up to us, we lowered our tow rope in the usual way so that theirs could pass over the top.

One of their horses, I noticed, was lame and Mary's father remarked upon this as they came by.

'Had a fall, has he?' asked Mr. Fisher.

'Aye,' came the reply. 'Two mile back and the next change is not for another two on this stretch. There's no certainty the animal will get there the way it's a-limping now.'

'Why don't you rest him for an hour or so,' Mary's father suggested. 'Might recover if it's only a strain.'

'No time for that, Captain,' said the man. 'We've a load of London goods for Runcorn. Has to catch a Liverpool boat in a week's time.' But then the man's eye was caught by my Jenny.

'That's a fine animal you've got there, Cap'n,' he said. 'How'd it be if you lent us her, as far as the next stage? We can pay you for her hire, and leave her for you to pick up at the stable. The lame 'un can come along with you then, in its own time.'

When he was told that Jenny belonged to me, the boatman

turned to me with a smile.

'Then 'tis you, young master, I must do business with,' he said. 'What do you say now?'

'You will pay the lad,' warned Mr. Fisher.

'Aye,' said the man. 'One shilling for the two mile pull, which is better pay than he'll ever see again. But we're in a fix with this lame nag, and that's a fact.'

'Pay him in advance then,' demanded Mary's father. 'So that he's sure of it.'

I was given a silver shilling—more money than I had ever seen in all my young life. Jenny was quickly harnessed into the place of the lame horse, in a matter of moments I was on her back and we were off.

It is strange how small things become big. I never for a moment supposed as I rode away that it would be the last I should see of Mary and the *Blackbird*—the last, that is, for several years. But that is to look too far ahead. I did not even bother to wave goodbye as I rode off, for I expected to see them again when they caught up with me at the staging post. Meanwhile, riding at a steady pace along the towpath, Jenny pulling well in harness with the other horse, I was able for the first time to watch a fly-boat being worked.

Everything was very different from the slow pace of a monkey boat. There were two locks to pass before we reached the staging post and the speed with which they were worked was a wonder to see. It was hurry, hurry, hurry. Get into the lock before that boat ahead. Fill it quickly as you can. Here you, boy, take this rope and hold it tight. Open the gates. Get the tow rope out to the horses. Off again; whip those lazy animals up. There's no time to lose. There's a ship to catch at Liverpool! I soon saw why they needed two crews, one off duty and one on. They worked at that pace, night and day, never stopping. Not even for a friendly jug of beer at a wayside inn. I could see that this was not the life for me.

Riding easily on Jenny, towing the fly-boat, I suddenly saw ahead the stables where the horses would be changed. Fresh horses were waiting on the towpath, already harnessed up, and as we approached, the ostler led them round until they were walking beside Jenny and her companion horse. Then, without the boat even having to slow down, the tow was unhooked and the new horses took over the pull.

It was neatly done and I looked with admiration as the fly-boat moved on, not delayed a moment by the change.

'Good luck, youngster!' shouted the fly-boat steerer, waving a friendly hand as the boat drew away.

I was unharnessing Jenny from the other horse when the stable master came on to the towpath. He looked curiously at me and then walked across to take a closer look.

'How come you're riding in my team—with this mare?' he asked. When I told him Jenny was my horse, explaining that his own lame animal would be arriving directly with the *Blackbird,* he frowned and pursed his lips.

'Maybe so,' he said. 'And then, maybe not.' Then he grabbed Jenny's bridle and half turning so that his voice could be heard in the yard behind the stables, he shouted:

'Hey, mister! I reckon this 'ere's your horse!' And to my horror, out from the yard came—Silas!

Jenny must have seen him too, for I had only to leap upon her back and shout 'hup!' when she reared like a circus horse, tearing her bridle from the stable master's hands, and we were away.

All the good food and care I had given her now began to tell. The mare tore down that towpath as if she was running in the Derby on Epsom Downs! We passed the fly-boat, my friend at the tiller watching me with amazement and his mate leading the newly harnessed horses making way for me to pass. We must have galloped for an hour. I could not be sure tha Silas was not following on another horse, and not until a goo twenty miles were covered did I at last slow Jenny down to walk. But even then, for another hour, I kept a very shar lookout behind.

Once more alone in the world with only a shilling in m pocket and Jenny to care for and feed, I now had to conside the best thing to do. It was summer time which was lucky fo that meant I could sleep in hayricks and barns whilst Jenn could find grazing on the canal banks and—if no farmers wer

about—in nearby fields. But I could only live like that for a few weeks at most. I still had to decide what to do after that.

One thing all boys could do in those days was run away to sea. My father had often told me about the great East Indiamen of a thousand tons and more, which traded to India and beyond. Of the smaller ships, frigates and schooners—and a few ships driven by steam engines too—which traded around the coast. There was always a place in one of those ships for an active lad that knew about boats and few questions were asked if a crew was short and the ship ready to sail. I decided therefore, to make for the coast, and Runcorn, on the Duke of Bridgewater Canal, was the place to go. There, I knew, cargoes from canal boats were transhipped to Liverpool. And in Liverpool Docks I would find ships sailing to every part of the world.

I did not set off right away. I had the sense to know that news passes quickly along the canals from boat to boat and if I was seen on the towpath, Silas would soon get to know. If I could find work away from the canal for a week or two, by then it

should be safe to continue on my way.

Once again, luck was on my side. I was riding Jenny slowly along the towpath beyond Birmingham when I came to a place where the canal widened into a basin with wharfs along one side. Several boats were moored there and gangs of men were loading coal. As I came up to them, one of the men saw me and called to one of his mates:

''Ere's the lad with th'orse, Bill.' Then he turned to me and pointed to where I saw some trucks with horses harnessed to them.

'Over there, lad,' he told me, 'and you've come just in time!'

The man seemed friendly enough and I could see that he had mistaken me for somebody else. So I went to where he pointed, ready to explain I was not the lad they were expecting. But I was not given the chance.

The foreman, for that was what Josiah Smutch later proved to be, gave me an impatient look.

'Get off that there horse, boy,' he ordered me. 'There's no time for fancy riding in these parts.' Then he turned to one of the other men who was holding some horses near the trucks.

'The lad's here, Dan,' he told him. 'Get 'im harnessed in there, and quick!' And before I knew what was happening, Jenny was harnessed to an empty coal truck and I was walking beside her, leading the poor bewildered animal to—I knew not where!

There were hundreds of these horse-railways along all the canals at this time. They were mostly short lines, built by collieries or ironworks to carry coal and iron to the canals for loading into boats. The railway I now found myself working on had a double line of rails and as I led Jenny along one rail, other trucks led by boys passed me down the other. One or two grinned in a friendly way as they went by, but most of the boys were pale and thin as though they were not properly fed. They looked unhappy and went along in silence without looking up. The horses too, were poor-looking and I could see that many of them had been put on the railway because they were too old for work on farms or along the canals.

I had already decided that this was not the right work for a fine mare like Jenny when we reached the end of the railway at the coal mine. There I could see a line of waiting trucks filled high with coal.

The foreman came out of his office hut as I approached.

'Sam Bonner's boy?' he asked me.

It would have been hard to explain how I came to be there. So I just nodded and said. 'Yes, sir.'

'Right,' said the foreman. 'Not 'afore time neither.' Then he pointed to the row of waiting trucks.

'Take the first 'un down. Josiah'll see you back.'

So I went to work for the colliery and was paid sixpence a day, which was good money for a boy my age. There was a stable where the trucks were kept and there Jenny and I spent each night; she standing up in her stall; myself asleep beside her in the straw. It was not the best way to live, but the other boys left me alone and I was certainly safe from the attentions of Silas Jones. The 'Sam Bonner's boy' for whom I had been mistaken, did turn up with his father's horse whilst I was there, and Josiah Smutch gave me a strange look once or twice as I changed trucks at the quayside. But he said nothing and at the end of two weeks when I drew my pay, I had twelve more

silver shillings to add to the single one with which I had begun.

I now decided the time had come to move on and with money enough to buy oats for Jenny and food for myself, the 70 mile journey along the towpath of the Trent and Mersey Canal was easy compared with the hard times I had just been through. It took us just three days to reach the Duke of Bridgewater Canal. We joined it at a place called Preston Brook and from there it was only a few more miles to Runcorn—and a ship to Liverpool. I would have to sell Jenny; that thought made me very sad but I would need money to buy seaman's clothes and other things for a life at sea. Jenny would fetch at least ten pounds, of that I was sure. But even so, as I sat on her broad, friendly back, it felt unnatural in some way, to sell such a faithful friend.

It was with these thoughts that I turned down the towpath of the Duke of Bridgewater Canal, to what I thought was to be the last short stretch of my journey to the sea. Then, once again, a chance happening changed the whole direction of my plans.

I was riding quietly along, admiring the long straight stretch of water ahead—for there were no locks on this canal from Manchester to the sea—when I suddenly heard from behind me the clatter of hooves and the shrill notes of a hunting horn. Turning quickly, I was astonished to see bearing down on me a team of three galloping horses ridden by lads no older than myself. Behind them in the canal was the strangest boat I had ever seen. It was planing along on top of the water, making hardly a bow wave, and the galloping horses were pulling it at what must have been ten or twelve miles an hour.

The boat was slimmer and altogether lighter in appearance than even the fly-boats and it was a positive racing skiff compared with *Blackbird* or *Swallow*. As it came swiftly towards me I could see passengers on board. Smartly dressed people; men with top hats and ladies in satin and lace. At the stern, the steerer was dressed in a uniform with bright brass buttons and a peaked cap. As grand a figure as any captain of an ocean-going ship!

I had little time to study this exciting scene for Jenny, frightened by the hunting horn, took to her heels and raced off down the towpath leaving the galloping team far behind. When I did at last stop her, she became so restive when she saw the other horses gallop by that only with great difficulty did I stop her from racing after them again. But the swiftly moving boat and its team were soon out of sight and I settled Jenny to a steady trot.

I was wondering to myself at what I had seen—for there were no passenger boats like this one on the Stratford Canal, nor on any others my family boat had travelled along. Then I saw a solitary horseman galloping towards me along the towpath ahead. At first I wondered if I might be in trouble for having got in the way. Or even worse, had somebody recognised Jenny as Grandad's horse—the steerer, perhaps or one of the boat crew? But when the horseman came up to me his first words quickly put my mind at rest.

'Captain Smith saw your gallop down the towpath, young man,' he told me. 'He likes the shape of you, and of your lively mare,' he nodded appreciatively at Jenny. 'The Captain reckons

you might both do for the swift-boat service if you've a mind to work on the canal.'

I was so taken aback that at first I could not find words.

'To ride—like those lads, in a galloping team?' I asked at last.

The man laughed. 'Aye, that's right!' he said. 'In the service of the Duke of Bridgewater Canal, and you'd have to go far for better work than that!'

So I did not run away to sea after all. Instead, I became a postillion—horseback rider, that is—in the crack swift-boat service on the Bridgewater Canal. It was an exciting life for we had to make the twenty-six mile journey from Manchester to Runcorn in time to catch the ebb tide for Liverpool and that meant getting there in not much more than two hours. There were stables along the canal, spaced at four mile intervals, so we had to change horses five times along the way. We postillions could make the change, jumping off the tired horses' backs and on to the fresh team in 30 seconds—flat! We wore bright scarlet coats with silver buttons, white breeches and black leather riding boots which we polished until they shone like glass. The lad who rode the leading horse had the hunting horn to blow. We took that in turns.

There were two towing ropes. The rope from the leading horse was fixed to a post in the front of the boat. The other two horses pulled a tow-rope fixed to the stern. The boat itself was seventy-two feet long and just wide enough for two people to sit opposite each other with room to pass between. Passengers could sit in the open at the front or else in first, second or third class compartments along the rest of the boat. These three compartments were covered in by a tarpaulin hood with small window openings along the side. In cold weather they were heated by a coal stove. Food and drink were served on board and although, of course, I never travelled in one of our boats myself, I was often told by passengers how much they enjoyed the trip. 'Much more comfortable than a bumpy stagecoach,' they said. And the way we rode those horses, just as quick.

I worked on the Bridgewater Canal for nearly ten years. The

canal manager bought Jenny from me for £25 and she was one of the horses kept in the Manchester stables to run the first stage of the canal. I lived with the other lads in a Manchester house owned by the canal company so I was able to see Jenny often, apart from riding her along the canal. There were grooms at the stable, of course, to look after her, but she was always pleased when I called in on my days off and took her a carrot or an apple to eat.

One day, I was riding lead postillion along the canal, sounding my horn from time to time to warn people out of the way. Along one stretch where we were always at full gallop, I saw an old monkey-boat ahead, crawling along at the usual snail-like pace. I gave a toot on the horn and saw the towing horse stop to let the tow rope go slack in the water in the usual way. Then the steerer moved his boat into the middle of the canal so that we could pass over the rope between him and the bank. We had to slow down slightly to pass the towing horse and as we came level with the boat, I saw the name on the side. It was the *Blackbird!*

A moment later I saw Mary at the cabin entrance. She was

grown to a young woman, and a very pretty one too. She saw me, but I was past and galloping away up the towpath so quickly that she had no time to recognise me from all those years ago. Directly I reached Runcorn, I borrowed a horse and rode back.

That is really the end of my story. A year after we met again, Mary became my wife and we borrowed Jenny to drive us to the church. Word of our marriage must have gone round the canals for my family got to hear of it and a few months later the *Polly* arrived in Manchester and I saw my mother and father again. Father told me that Grandpa never did buy another horse when I stole his. When Silas Jones went off to look for me, he never went back. Left alone, Grandpa decided he was too old to work any more on the canals. He lived on the *Swallow*, moored at Stratford-upon-Avon, until he died. He left the *Swallow* to me in his will!

Mother reckoned he had a guilty conscience about the bad way I was treated and knew that if he had not got drunk himself that night, there would never have been any need for me to run away. He left me the boat to make up for what he had done.

So it all turned out for the best in the end. Mary and I are old now and have lived on the *Swallow* for many years. The old monkey-boat looks fine, painted up with pictures of castles, garlands of flowers and moons and stars. Oh, and one last thing. She is no longer called the *Swallow*. We renamed her the *Jenny*. For if it had not been for that faithful old friend, Mary and I would probably not have met again. And this story would never have been told.

2 The Inside Story

Canals are roads paved with water. Ordinary roads need to be surfaced with hard-wearing materials like concrete or stone but a water road makes its own surface. Pour it in, let it settle, and it will be perfectly smooth. Keep it supplied with water and it will never wear out.

Rivers, or course, were the first inland water roads and have always been used to carry passengers and goods. But they are really big drains, collecting rainwater from the land, draining it into the sea, twisting and turning across the countryside on the way. It is nearly fifty miles along the river Thames from Windsor to London Bridge, but only twenty-five miles by road. Rivers can be improved, widened, deepened and straightened to some extent. But they have to follow the valleys and can never be as straight as roads.

When, therefore, water roads were needed to join places where no rivers ran, canals were built. They were made as straight as the countryside allowed and for nearly 100 years they were busy with goods and passengers. They linked towns to towns and ports with cities inland.

Now we come to the Inside Story of how they were built, how they were used and how, in the end, they were overtaken by an even better kind of road—the railroad.

BUILDING A CANAL

Anyone who has played on the sands, making waterways when the tide is out, will know how difficult water is to control. It leaks through carefully made embankments, drains away into underground holes, or overflows without warning when the tide comes in! But most difficult of all, it will only stand still if its bed is level. Tilt it only a very small amount and it will immediately rush downhill. To be successful, therefore, canals must run straight and level—straight, for the shortest distance between two points; level, to stop the water from running away.

To be useful, moreover, canals had to run from town to town for the cargoes they carried were needed by factories and shops. Yet no two towns in Britain are at exactly the same height above sea level. Some are exactly at sea level. Some are in valleys which lead to the sea. Others are on top of hills. The land between them is hardly ever flat. Canals joining towns must therefore be able to climb hills.

This was done by building a 'staircase' of locks up the side of a hill, like this.

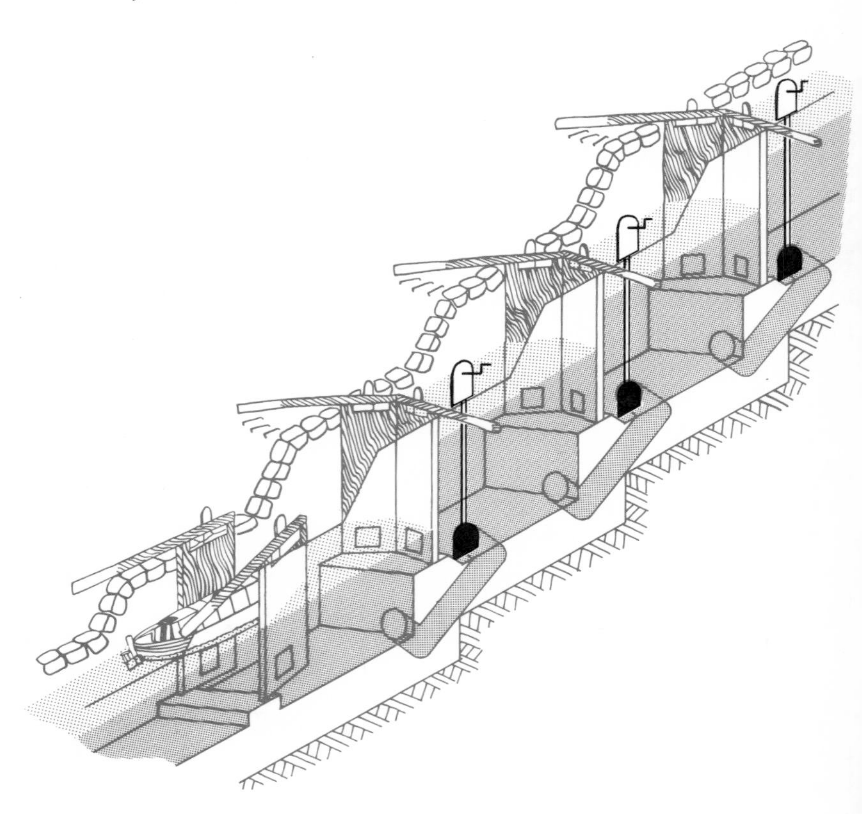

The locks lead into each other and although a boat can go *down* the staircase on one lock-full of water, in order to *climb* the stairs, each lock in turn has to be filled except the bottom one. Work it out for yourself by studying the picture above.

Water is let into a lock either by means of underground channels which can be opened by means of 'paddles' which can be raised and lowered by winding handles.

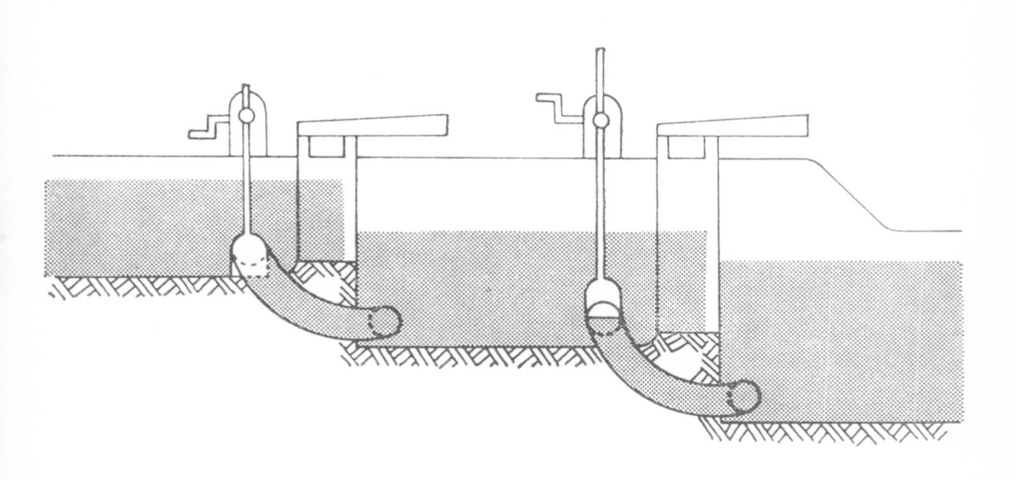

Or by paddles in the gates themselves. The water is almost always let out of locks by gate paddles.

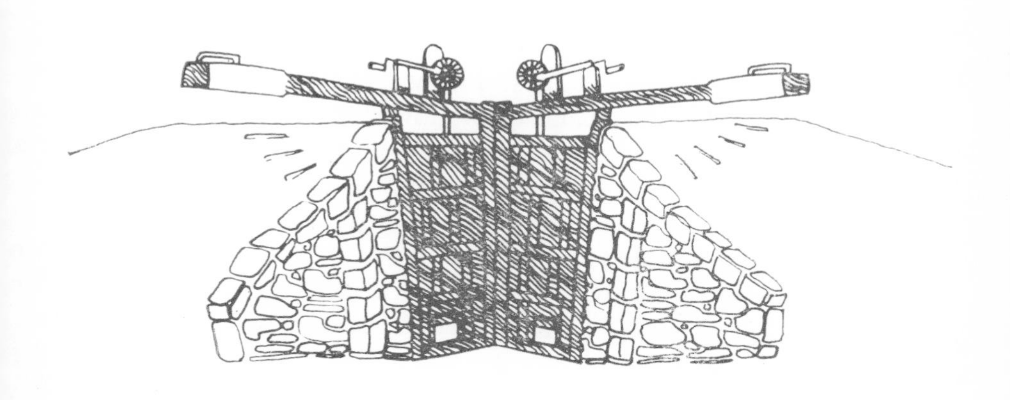

This is how a complete lock looks. Notice that the gates close together into the direction the water flows. The pressure of water, by this means, helps to close them tight.

WATER SUPPLY

Locks are hydraulic lifts. The water they use flows from the higher levels of a canal to the lower levels and eventually into the sea. If a canal has a lot of locks along its course, it will use a great deal of water. A narrow lock uses about 25,000 gallons when it is filled and emptied (enough to fill a big swimming pool) and a wide lock more than twice as much. All this water has to be replaced, otherwise the canal would run dry. Water has therefore to be poured into the canal at its highest point and also at places along its course. This is done by collecting water in reservoirs either by damming streams or by pumping water up from rivers at lower levels. The supply from the reservoir to the canal can then be controlled to replace the water used in the locks. If there is a lot of traffic the need for water will be great and in summer, if the reservoirs fall in level because there is no rain, a canal might even be closed to traffic because there is insufficient water to work the locks.

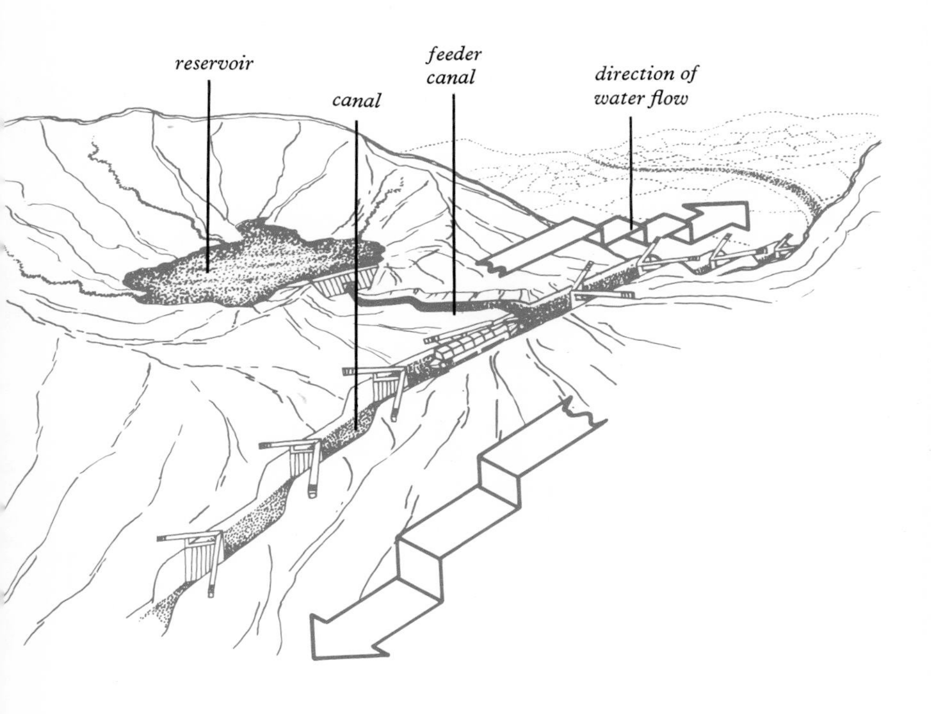

At times of little traffic and plenty of rain, water might overflow the banks if arrangements were not made to prevent it. Overflow weirs are provided along all canals for surplus water to drain away, usually into nearby rivers. Canal water is often used by the towns through which they pass, for industries and to add to the local drinking water supply.

STRAIGHT AND LEVEL

The canal builders had to solve many problems. In the picture below you can see how they had to build bridges, aqueducts, cuttings and tunnels as well as locks, in order to keep their water 'roads' straight and level.

Tunnels were very expensive to build so they were made as narrow as possible. This meant that only one or two of the shortest tunnels had towpaths inside for horses. The most usual

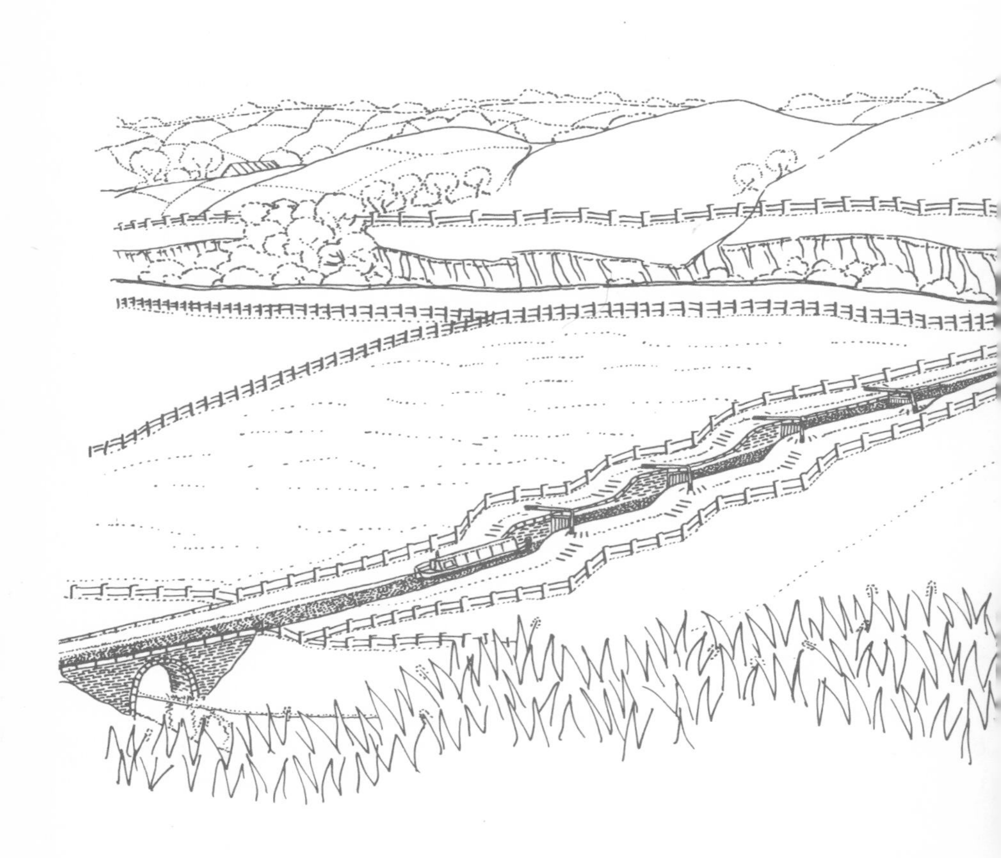

way for boats to be taken through tunnels was by 'leggers' who walked the boats along. Boards were fixed to the front of the boats, one on each side, projecting over the water. Men could then lie on these and, holding on to the boards, could walk themselves — and the boat — along the tunnel walls.

Leggers waited at the entrances of most tunnels and could be hired by boat captains to take their boat through. Their usual charge was one shilling and sixpence (about 8p), for the trip, which was not too much for a tunnel like that at Standedge which is over three miles long. That trip must have taken nearly 3 hours and all that time the leggers were in absolute darkness with water dripping on to them from the tunnel roof.

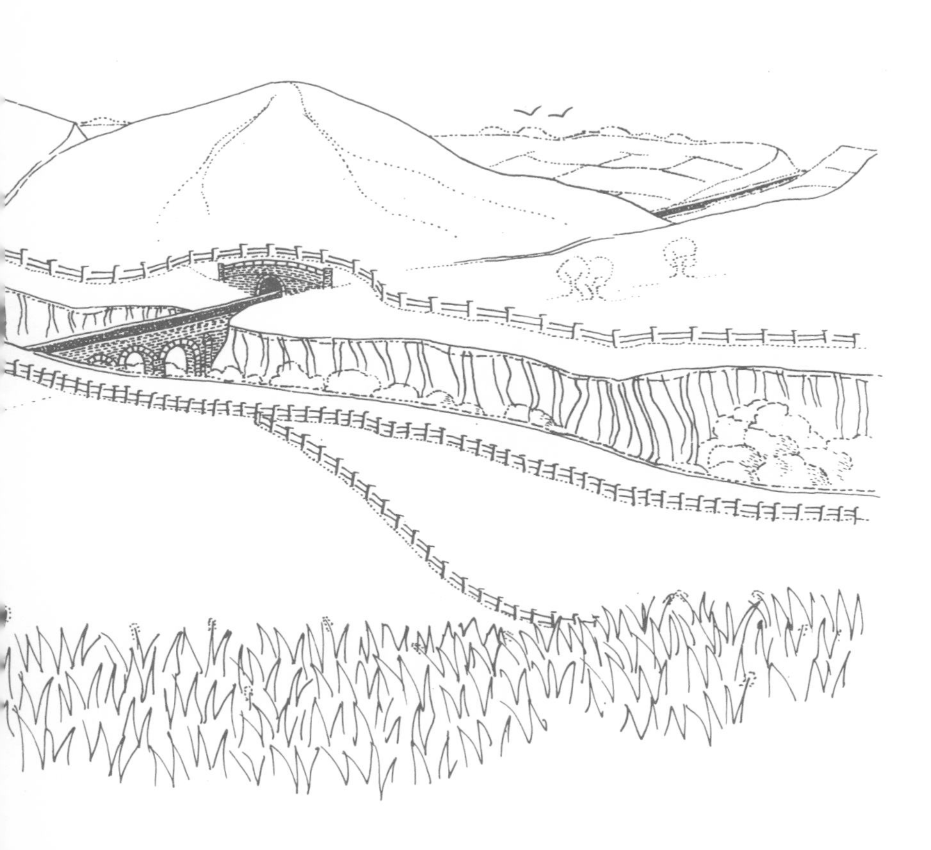

CANAL BOATS

In the story, you read about three kinds of canal boat. There was the monkey-boat or narrow boat:

This was the maid-of-all-work on the canals and its narrow shape was simply a way of avoiding the need for a wide canal. A long, narrow boat is, of course, very 'unseaworthy'. In rough water such a boat will quickly capsize and sink. But in the perfectly smooth water of the canals they were quite safe — unless they were badly overloaded. Then, they often overturned and sank. Narrow boats were built to a pattern and were nearly all seventy feet long, seven feet wide and had a draught in the water of only 8 inches when they were empty. As they were loaded, they sank down into the water by about one inch to

each ton loaded and the maximum load was thirty tons. Fully laden, therefore, they sat down in the water almost three feet. Most of the canals were made four feet deep, so the water supply could not be allowed to dry up or, in no time, every boat on a canal would be aground!

The narrow boats were often used as homes. The tiny cabin was only about eight feet long by five feet high and six feet wide. Three children under twelve and their mother and father were the most that were allowed to sleep in that small space by law. But even that size family must have found the space very cramped.

The fly-boats which you read about in the story were simply smaller and lighter versions of the narrow boats which carried lighter cargoes needed in a hurry.

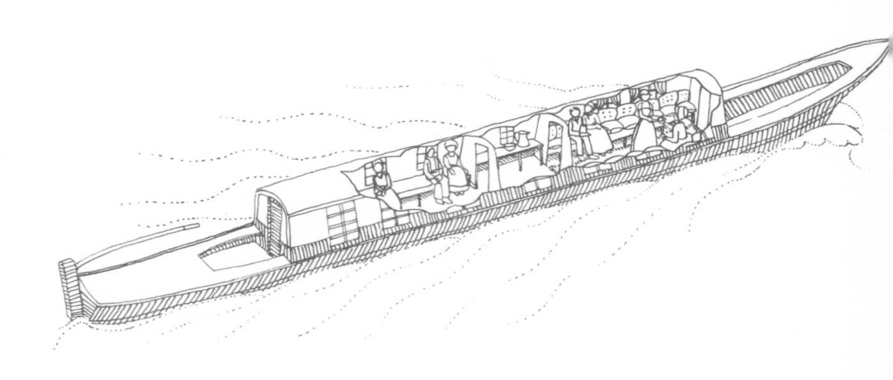

The swift boats, which Jenny and her friends towed along the Bridgewater Canal, were specially built. They were able to travel so fast because they were built to be actually lifted by the power of the towing horses to plane on their own bow wave. That is why they needed two or three very strong horses not to pull them along so much as to exert that first powerfu effort needed to lift them on to the top of the wave. From then on they could skim along on top of the water, making hardly a wash—and this was important because a big bow wave rushing along would have caused too much damage to the banks of canals.

These express boats were very comfortable. They were about the same length as the narrow boats, about 70-72 feet, but they were only five feet wide and had a draught of no more than three inches when they were loaded with as many as seventy passengers. They must have been quite a sight to see, planing down the canals at twelve to fifteen miles an hour!

HOW IT ENDED

The canals were busiest from about 1790 to 1840. During that time the length of inland waterways and navigable rivers rose from 2,223 to 4,000 miles. The canals were extremely busy. In 1821, for example, a gentleman named Josiah Baxendale took his son to inspect some of Pickford's depots along the same stretch of canal near Birmingham which young Will Rowlands rode along on Jenny. They passed more than 400 canal boats in a six-and-a-half-hour journey.

Pickford's, the same firm that now has warehouses for storage and carries our furniture removals in vans, were the biggest owners of canal boats. In 1838 they employed 500 men and boys on their fleet of narrow boats.

There were more than 3,500 miles of railway, all worked by horses to serve the canals and the stagecoaches were finding competition from the swift boats taking their passenger trade away.

Then came George Stephenson and the Puffing Billy. Steam engines took the place of horses on the railways and they were soon pulling trains filled with the goods which were once carried on the canals. For a time canals and railways worked side by side in competition, but one by one, the railways bought up the canal-owning companies and closed them down. Canal after canal went out of use.

NEW TIMES

The canals of Britain have now begun a new lease of life. Instead of heavily laden narrow boats, the waterways are now busy again with pleasure craft of all kinds. There are only 2,000 miles of inland waterways left but these are now used by holiday makers. The locks are busy once more; so are the canalside pubs. If you know where to look, you can still see the stables where the horses were kept for staging the fly-boats; the marks on the walls of tunnels made by the leggers as their heavy boots pushed against the brickwork can still be seen; warehouses still stand alongside the canals in towns which were once stocked with goods brought by boats. More and more derelict canals are being brought back into use by people who have formed themselves into Societies and work voluntarily in their spare time. Why not see if there is work to be done on a canal near you?

MAP OF CANALS AND NAVIGABLE RIVERS STILL IN USE
Canals
Navigable Rivers

3 Where to see Canals

If you look at the map on page 39 you will see just how many canals and navigable rivers are still in use in Britain. A journey along one is like visiting a great open-air museum. Along the banks can be seen the factories, mills, lock-keepers' cottages, bridges and inns which served the canals in their heyday. On quiet stretches it is easy to imagine the canal full of the sights and sounds of the great days of canal transport – the creak of tow ropes and the steady clip-clop of the horses on the tow-paths. Canals and rivers often take you through parts of the countryside that you would never see any other way. Even places you know well can look completely different when seen from a canal. Wild life of all kinds thrive along their sides. When you start to explore you will quickly find there are many things to be enjoyed on the canals besides their history.

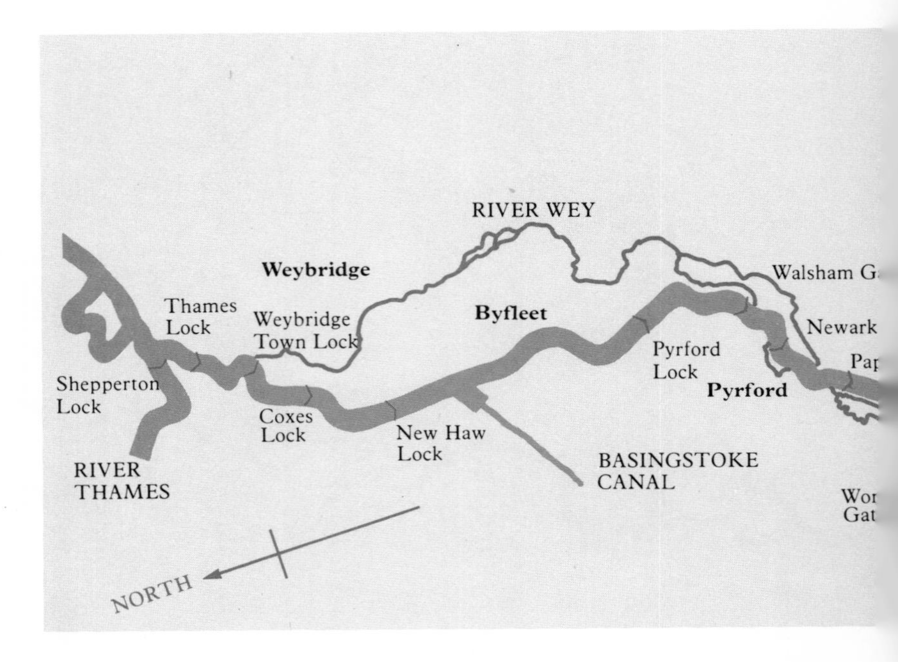

The National Trust owns part of two canals as well as one river navigation. These are described in the list that follows together with some museums where you can see exhibits about canals and their history.

KENT
Royal Military Canal, nr Appledore
The National Trust owns three and a half miles of this canal. Originally it was built as part of the defences on the south coast against Napoleon's plan to invade England. Later it was opened for commercial boats to use but never became an important waterway. Today it is a pleasant tree-lined canal open only to rowing boats and anglers.

SURREY
River Wey Navigation
This is a fine example of a river which has been made navigable by building locks and canal short-cuts along its length. The work was completed as far as Guildford as early as 1653. In

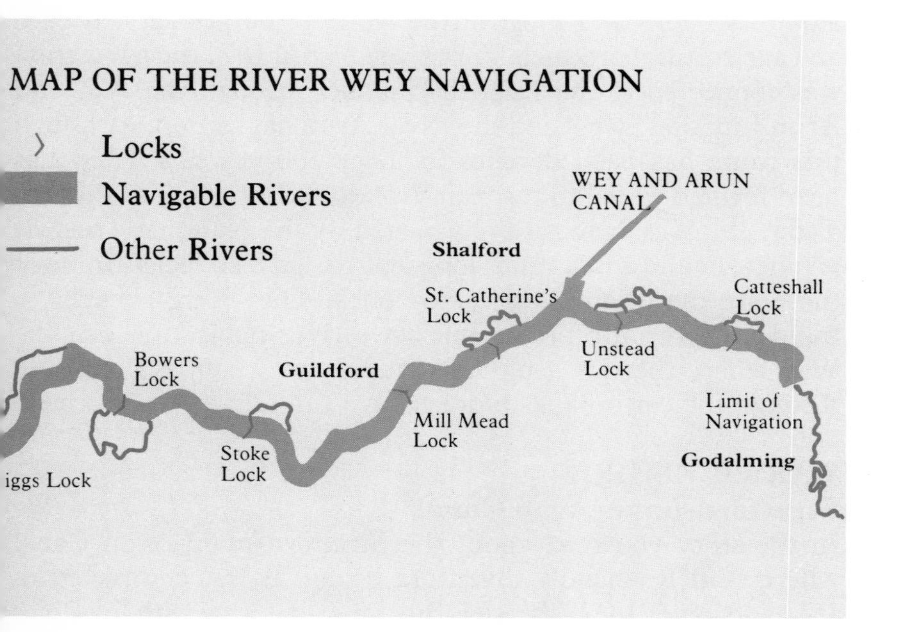

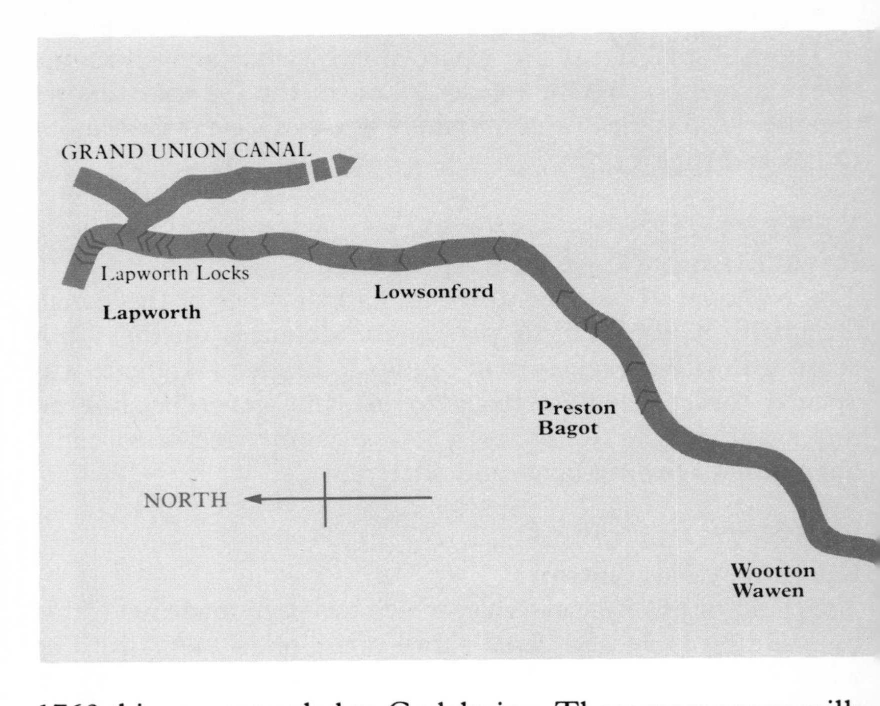

1760 this was extended to Godalming. There were many mills along its banks and most of the working barges carried corn and agricultural produce. For 55 years until 1871 the navigation was connected to the English Channel through the Wey and Arun Junction Canal and the River Arun navigation. Although this route has been derelict for over 100 years, a Trust has been formed to try to re-open 'London's lost route to the sea'. Today the Wey Navigation is owned by the National Trust. It is nineteen and a half miles long and has sixteen locks. Some of these are the original turf-sided ones. They were cheaper to build but are only practicable on a river navigation like the Wey where there is a plentiful supply of water as it escapes through the earth banks more easily.

WARWICKSHIRE

Stratford-upon-Avon Canal

In the story you read about the Stratford-upon-Avon Canal where Will Rowland's adventure began. It was completed in 1816 and joined the Worcester-Birmingham Canal with the River

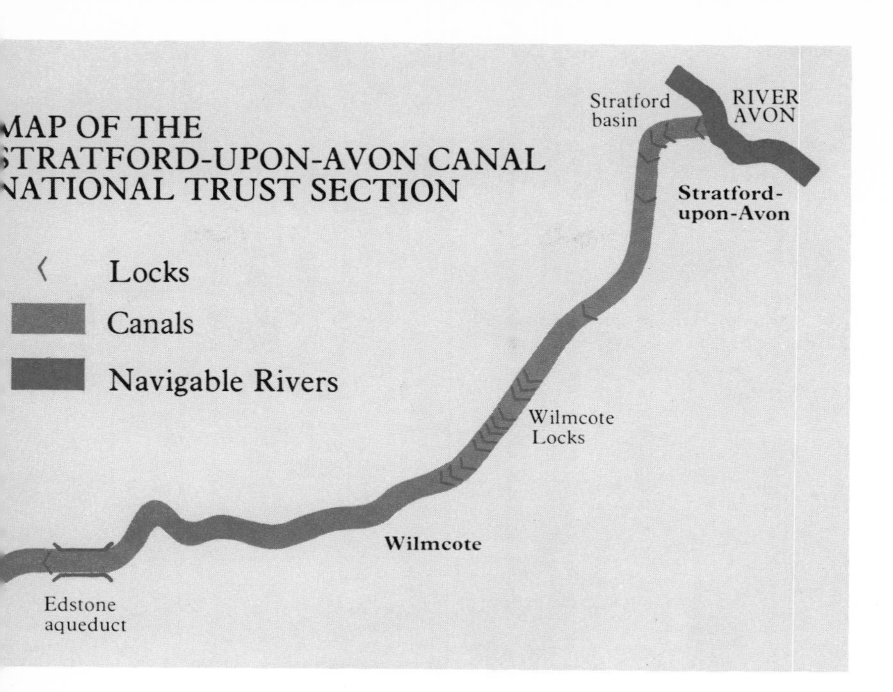

Avon. Boats like the *Polly* carried coal from the Midlands, returning with corn from the farms around Stratford. Like all the canals, competition from the railways was too great and by 1960 the southern section was derelict. It was then that the National Trust acquired the thirteen and a half miles of the southern section. With the help of volunteers the canal was cleared and re-opened in 1964 showing the way for many more successful schemes to rescue canals. There are 56 locks along the canal and 36 of them are in the National Trust section. There are also three aqueducts and twenty-six cast-iron bridges.

WEST YORKSHIRE

Marsden Moor

Along the top of Close Moss on Marsden Moor (owned by the National Trust) can be seen the line of air-shafts which supply Standedge tunnel 600 feet below on the Huddersfield Narrow Canal. This is the longest canal tunnel in the country. It is more than three miles in length and used to take three and

River Wey Navigation

a half hours for the boats to be 'legged' through. It has been closed since 1951. The air-shafts which ventilate the tunnel were originally used for digging it out. By digging from the bottom of the shafts the builders were able to tunnel from several different points instead of just each end. This meant more men could be used and the work completed more quickly.

MUSEUMS TO VISIT

The Boat Museum, Ellesmere Port, Cheshire
The former Shropshire Canal buildings are used by the museum, and an old tollhouse contains small exhibits. There are twenty boats to see in the basin all with a tale to tell of the canals and the folk who used to live and work on them. The museum aims to preserve not only the boats but the skills and techniques

that went into building and repairing them. Visitors can watch restoration work in progress.

The Canal Museum, Dewsbury, West Yorkshire
Displays on the Calder and Hebble navigations as well as items of general interest to do with canals.

Corinium Museum, Cirencester, Gloucestershire
This museum is mainly concerned with displays of Roman Britain, but an interesting exhibition dealing with the Thames and Severn Canal has recently been added.

Crofton Beam Engines, Great Bedwyn, Wiltshire
Here you can see a pair of steam-driven beam engines built to pump water to the top level of the Kennet and Avon Canal. They have recently been restored and are steamed on certain weekends during the year.

Exeter Maritime Museum, Exeter, Devon
The museum is housed in warehouses around the basin of the Exeter Ship Canal. Among many exhibits are an early steam drag-dredger from the Bridgwater and Taunton Canal and a

Stratford-upon-Avon Canal

tub-boat which was fitted with wheels for travelling up the inclined planes on the Bude Canal.

Flint Mill, Cheddleton, Staffordshire
Two water-mills are preserved here in working order. The museum has displays on canal transport and the *Vienna,* a restored 70ft horse-drawn narrow boat, is moored on the Caldon Canal beside the Mill.

Goole Museum, Goole, Humberside
In the museum you can find items relating to the waterways near the docks.

Ironbridge Gorge Museum, Telford, Salop
This is now a major open-air museum of the Industrial Revolution. Here you can see part of the Shropshire Canal including an inclined plane which once served it together with the tub-boats which were hauled up and down.

Leicester Museum of Technology, Leicester, Leicestershire
There are several exhibits concerning canals in the museum.

Llangollen Canal Exhibition, Llangollen Wharf, Clwyd, Wales
An exciting exhibition of canals and their story. There are trips along the nearby canal in a horse-drawn canal boat.

Lound Hall Mining Museum, Lound, Nottinghamshire
Included in this museum about mining is a display on underground canal barges which were used in some mines.

Manchester Museum, Manchester
The museum includes an interesting canal display including a model of a canal ice-breaker *North Star.*

National Maritime Museum, Greenwich, London SE10
There are several items on canals in the museum. Particularly interesting is the 53ft Thames steam launch *Donola* and the steam tug from the Manchester Ship Canal.

The Science Museum, South Kensington, London
In the transport section there are several models of canal craft and interesting displays on the development of the inland waterways.

South Yorkshire Industrial Museum, Doncaster, South Yorkshire
In the Waterways Room you can see several exhibits on canals.

Waterways Museum, Stoke Bruerne, nr Towcester, Northamptonshire
This is the most important museum on inland waterways in the country. It is housed in a converted corn-mill near a flight of locks on the Grand Union Canal between Northampton and Towcester. There are items illustrating every aspect of canals and canal life. You can see the harness used by canal boat horses, costumes, boat engines, models, and much more. Particularly interesting is the reconstructed cabin interior of a typical narrow boat such as Will Rowlands in the story might have known.

MAP OF PLACES
MENTIONED IN CHAPTER 3
Canals
Canal Museums
National Trust sites
Dewsbury
Goole
Marsden
Moor
Doncaster
Manchester
Lound
Ellesmere Port
Cheddleton
Llangollen Wharf
Telford
Leicester
Stratford-
upon-Avon
Canal
Stoke Bruerne
Cirencester
Science Museum
Great Bedwyn
Greenwich
River Wey
Navigation
Royal
Military
Canal
Exeter